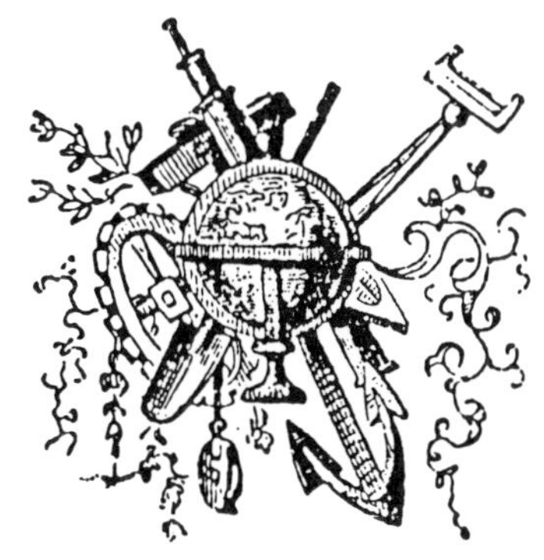

THE HIGHWAY
OF ALL NATIONS

THE HIGHWAY

OF

ALL NATIONS

BY JOSEPH FOOS.

Columbus:
PRINTED BY P. H. OLMSTED.
1830.

THE
HIGHWAY
OF ALL
NATIONS

BY

JOSEPH FOOS

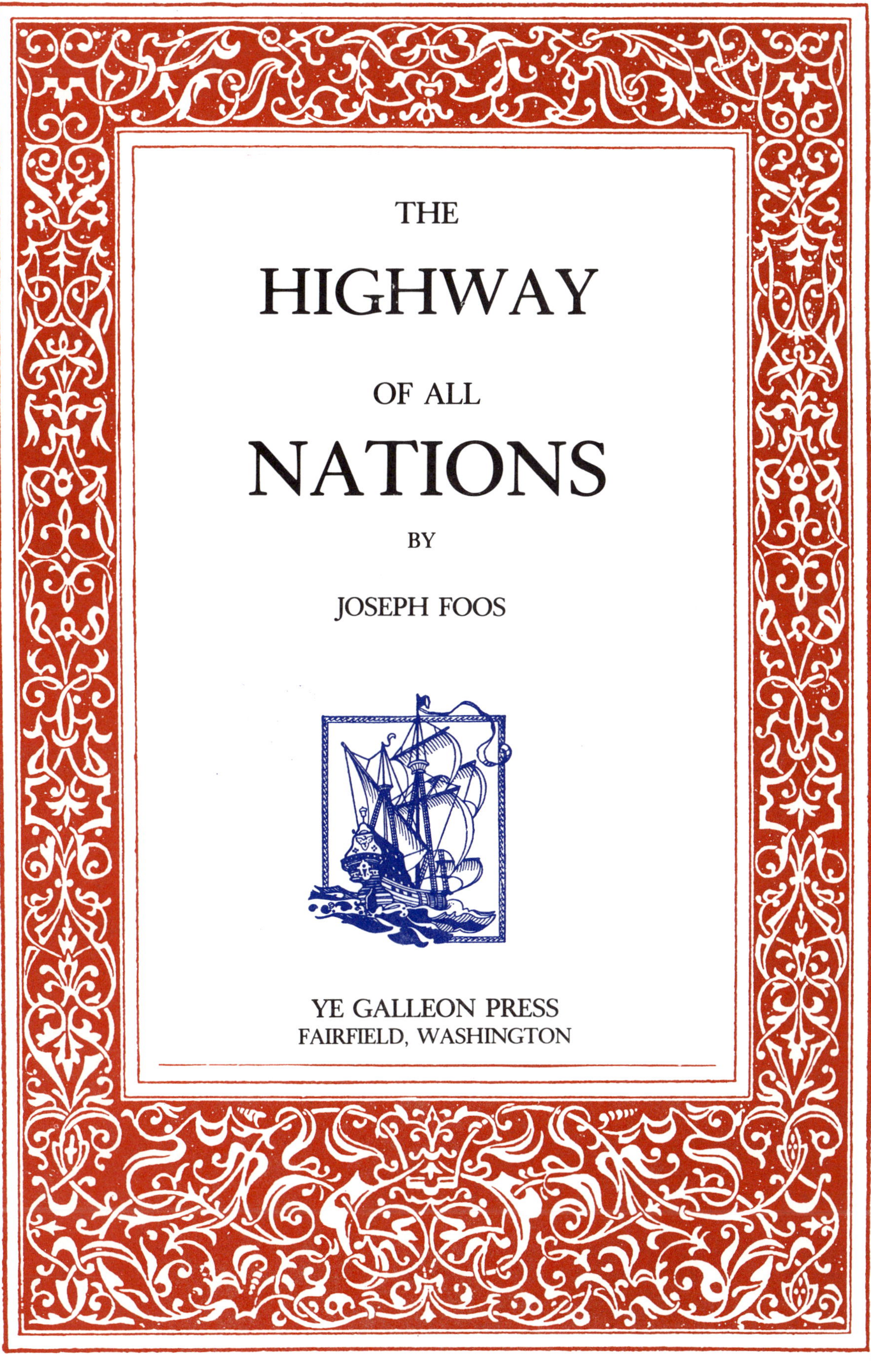

YE GALLEON PRESS
FAIRFIELD, WASHINGTON

Library of Congress Cataloging-in-Publication Data

Foos, Joseph.
   The highway of all nations.

   Originally published in 1820.
   Includes index.
   1. Canals, Interoceanic — Panama — Panama, Isthmus. 2. Canals,
Interoceanic — Nicaragua. I. Title.

HE532.F66   1987   386'.445   87-25344
ISBN 0-87770-421-X

TABLE OF CONTENTS

Joseph Foos was a State Senator in the Ohio legislature for several terms before and after the 1820 date of the printing of this pamphlet. I was unable to find biographical information on him or a portrait. Photography did not exist in 1820 but painting and engravings did.

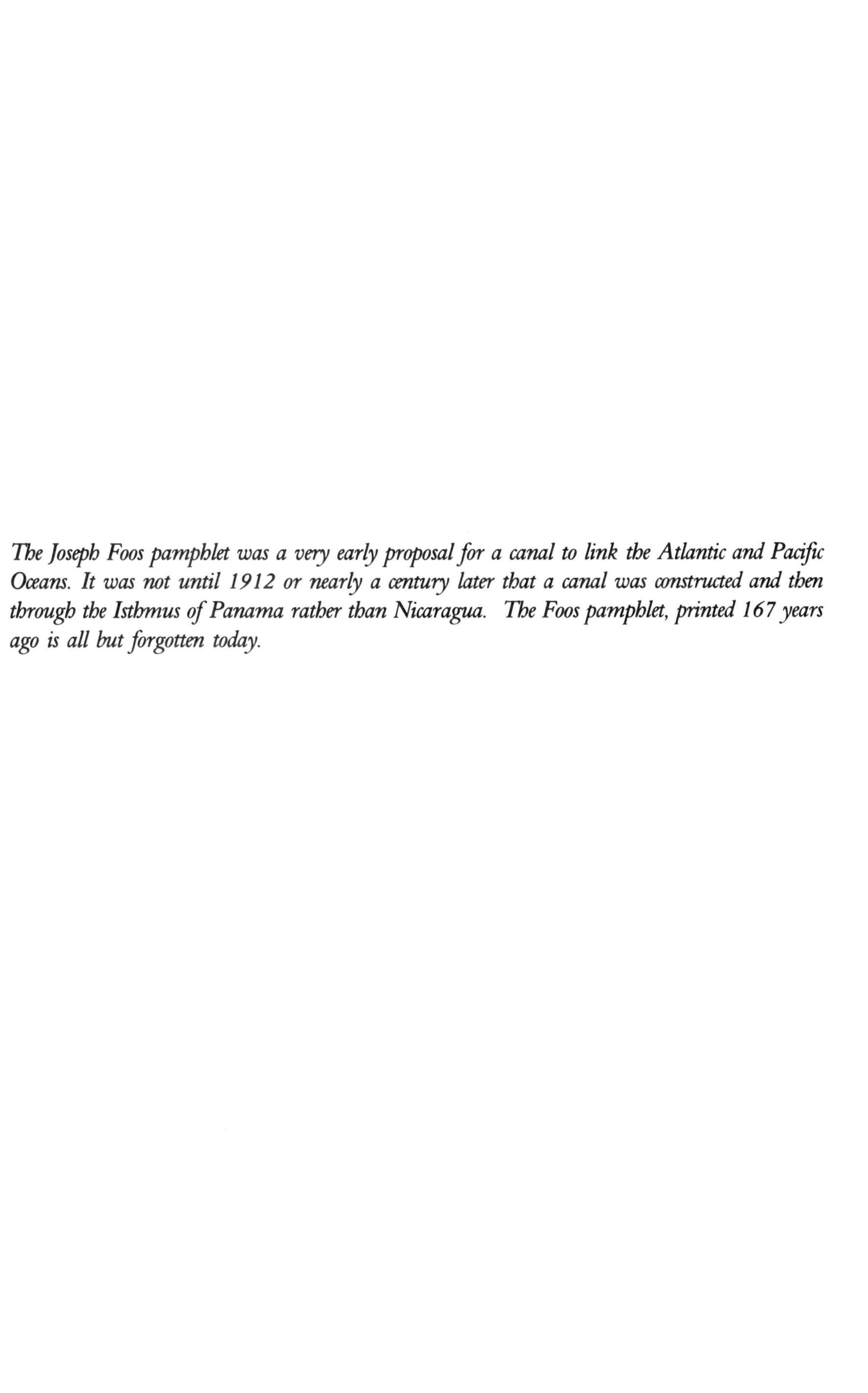

The Joseph Foos pamphlet was a very early proposal for a canal to link the Atlantic and Pacific Oceans. It was not until 1912 or nearly a century later that a canal was constructed and then through the Isthmus of Panama rather than Nicaragua. The Foos pamphlet, printed 167 years ago is all but forgotten today.

PREFACE

HE PUBLICATION OF THE FOLLOWING SHEETS ORIGIN- ated from a preamble and resolution, that the author of- fered in the Senate of the state of Ohio, on the 26th of January, 1819, which was laid on the table for consideration; but, on motion for its passage, was lost. It was entered on the journals of the Senate, and published by the Editors of the *Columbus Gazette* and the *Ohio Monitor*, soon after which it went on the rounds through the southern and eastern papers, from New-Orleans, via New-York and Boston, to Montreal in Canada. From the last, to wit, the *"Montreal Star,"* the Editor of the *Ohio Monitor* makes an extract, to which he prefixes some remarks, and calls on the author to explain to him and the public the prominent features and points contained in said preamble and resolution: all of which are here- unto subjoined.

The great difficulty that arises in treating of this subject, is the want of correct maps, charts, and actual surveys of the several points under consideration. Those that do exist, are in possession of the Spanish Government, consequently not accessible to an American citizen. The author has been for several years endeavor- ing to collect the necessary information; but from his interior situa- tion, little could be obtained, though he has sent to the book stores in the Atlantic cities.

The reason why the territory in question is so little known, is owing to the jealous fears of the Spanish government. But may we not indulge the hope that despotism will shortly be driven from this hemisphere, and be compelled to take its flight towards the con- taminated precincts of the old world, from whence it came: then will the American continent be left free.

Notwithstanding the difficulty in procuring the necessary materials for the following work, the author flatters himself that he will be able to furnish sufficient to justify a belief of the practicability of the project.

The author solicits the indulgence of a generous public, and requests that the work may not be condemned, without a full investigation of its merits, and that by a competent judge.

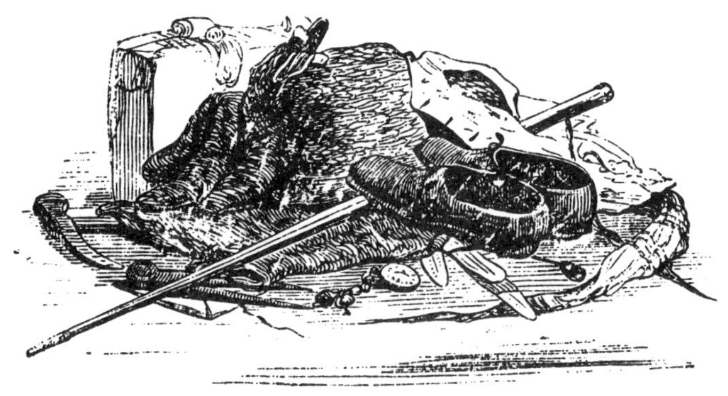

HIGHWAY OF ALL NATIONS

Extract from the Journal of the Senate, January 26th, 1819.

MR. FOOS MOVED THE ADOPTION OF THE FOLLOWING resolution:

WHEREAS, it would be of the greatest importance, in a commercial point of view, to every part of the United States, and especially to the states and territories bordering on the Mississippi, and the waters leading thereunto, that an easy and direct communication should, if practicable be opened across the narrow part of the American continent, from the *Spanish Main*, to the *Pacific Ocean*, for the following reasons, to wit:

1st.— The long and dangerous passage around Cape Horn, could be avoided; the distance to the North-west coast, and the East-Indies, would be shortened ten thousand miles; calculating from the 12th degree of north latitude, the point thought to be the most practicable, to the 56th of south latitude, the place of doubling Cape Horn, then returning north in the Pacific, to the aforesaid latitude of 12 north.

2d.— It would afford a short and easy communication with that part of the north-west coast claimed by the United States. It would completely lay open to the use and control of our government the whale and seal fisheries in those seas, and the fur trade on that coast, with the facility of planting and protecting commercial colonies in those immense and fertile regions of the west. The American empire, and the commercial enterprise of a free people, could and would be extended as circumstances might require. Our fellow-citizens would be enabled to participate deeply in the rich traffic of the East-Indies; especially as that trade could be carried on by the whale, seal and furs found on the north-west coast, thereby saving our specie from continual exportation to a quarter of the world that has heretofore swallowed up the precious metals of every age and nation.

3d.— The great advantage resulting to the western states and territories, in the completion of this project, would be the proximity of the canal to the mouth of the Mississippi, and the facility of ascending that noble stream and its great navigable branches, with steam boats to our doors. In this event, New-Orleans, or some point further down, would become one of the greatest commercial depots in the world, and secure to the people of the western country a sure and lasting market for their exports. Therefore,

Resolved by the General Assembly of the State of Ohio, That our senators and representatives in Congress be requested to make application to, and use their best endeavors with, the general government, to apply to the Court of Madrid for the privilege of examining the ground and opening a canal for the passage of large vessels from the Spanish Main, across the continent at Lake Nicuragua, or such other point in that quarter as may be found most practicable.

Resolved, That his excellency the Governor of this state be requested to transmit copies of this preamble and resolution to the president of the United States, the vice-president, the speaker of the house of representatives, and to each of our senators and representatives in Congress: And also to the executives of the different states and territories, with a request to lay the same before the legislative bodies over which they preside.

———•———

FROM THE OHIO MONITOR

Among other subjects, that have excited the jealousy of British prints, we find by the following lines from the *Montreal Star*, that there is considerable excitement produced by the bare suggestion that our nation might one day construct a canal to unite the waters of the Pacific and Atlantic oceans. And the fearful anticipations of the writer have carried him far in advance of the fact. For he states that the state of Ohio has petitioned the government of the Union to apply to the court of Madrid. Such a resolution was introduced into the Senate of this state, by General Joseph Foos of this country, but failed. The inducements to this undertaking are well set forth in the preamble to his resolution, offered last winter; but the feasibility of constructing such a canal is somewhat more doubtful. But as we have never made this a subject of much investigation, we hope the general will inform us, and the public, through the medium of our paper, of the practicable points, for a canal, the contemplated distance, and such other information as will best enable the public to judge of the expediency of the measure proposed.

———•———

FROM THE MONTREAL STAR

We have lately seen the solicitude of the United States of America for territorial aggrandizement: the great care and mathematical precision with which she lays down her boundary line between her territory and that of the Canadas; the rigorous measures adopted by the Governors of the Missouri territory, to extend and perpetuate fur trade: the anxiety she has invariably exhibited, for the prosperity of her Columbian establishment on the Pacific ocean: her right to fish on the coast of Newfoundland and Labrador in common with every British subject: her late valuable acquisition of East and West Florida, and the recent petition of the state of Ohio, to the general government of the Union to apply to the court of Madrid for a tract of land, a few hundred miles in length, between the Gulf of Darien and the Bay of Panama, in order to cut a canal for their exclusive use: we hesitate not to avow it as our opinion, that, the states of America are, at this moment, making secret overtures, to secure to themselves, that invaluable portion of territory: nor do we think, that the venal government of Spain, which is now a state of dilapation, will be large in her demands.

We have often been led to admire the secret policy of our American neighbors, but we candidly confess, that we never till now, had a proper conception of the full extent of their political wisdom. While the court of England is wasting in Eastern indolence; while that of France, is daily becoming more obnoxious to the people; while that of Spain, at once the terror and the curse of the subject, is rapidly becoming effete; the Eagle of the United States, is spreading its wings with speed, over the continent of North America. From a government surcharged with perfidy as Spain, what has England to expect; the pretended losses of the United States' citizens from the piratical baubles of Spain, may serve very well to hoodwink John Bull, till the secret emissaries of Monroe have finally concluded with the tools of Ferdinand, their overtures of negociation.

Those who doubt the capacity of the United States, or rather any single state of the Union, to effect a conjunction of the Atlantic and Pacific, by a canal navigation, have only to turn their eyes to the great western canal of the state of New-York and be convinced, that an inland navigation of three hundred and sixty miles in length, can be undertaken by a single state. Surely the objects of this Western Canal, in a political, or commercial point of view can never, by any rational being, be named in competition with a canal which must inevitably command the greater proportion of the western commerce of all North and South America, and much of the East Indies, as well as that of all the islands thickly scattered throughout the extensive North and South Pacific ocean, Russia, Tartary, &c. When we take a retrospect of what might have been done by our country a century ago; when we learn that Mr. W.

Patterson, a learned and intelligent Scotchman, applied to the government of Great Britain, to cut a canal, and to form an establishment at the isthmus of Darien, upwards of a century ago, what true British subject; what man that wishes well to his country, but must deprecate that worthless system of policy that defeated such a noble design.

I have observed, in a late number of the *Ohio Monitor*, an extract from the *Montreal Star*, in which several political and commercial subjects, relative to the United States, are briefly noticed, but with sufficient warmth and energy to show the jealous interest the British editors take in observing the rapid march of our national improvements. Amongst the subjects alluded to in the extract, is the *project* I had the honor to submit, in a preamble and resolution, to the consideration of the senate of this state, last session, requesting the general government to apply to the Court of Madrid for permission to examine the ground and construct a *canal* from the Spanish Main through Lake *Nicaragua*, across the American continent, to the *Pacific Ocean*. In the same number, the editor of the *Ohio Monitor* prefixes some remarks on the above subject, and calls on me to inform "him and the public, through the medium of his paper, of the practicable points of a *canal*, the contemplated distance, and such other information as will best enable the public to judge of the expediency of the measure proposed."

For me to attempt a full investigation of this important subject at present, is more than the pressure of business will admit of, and more than I ever shall be able to effect.

But as the object of enquiry is a laudable one, which deeply involves the commercial interest of our country, I will, with pleasure, from time to time, communicate such information as I may possess on that interesting subject.

The investigation of which will involve the following inquiries into a number of natural causes and effects, most of which are either remotely or proximately [*sic*] concerned wilth the project under consideration.

1. Such as the breadth of the continent at the points to be examined.

2. The elevation of the ground and distance to cut.

3. The pressure of the *Trade winds* in that quarter.

4. The accumulation of the water in the *Spanish Main*, and *Gulf of Mexico.*

5. A similar accumulation of water in the straits of *Babelmandel* and the *Red Sea*, produced by the same cause.

6. The origin of the *Gulf Stream*, its circuit, celerity and termination.

7. The difference in the elevation of the water in the *Spanish Main* and the opposite *Ocean*, the cause thereof.

8. The difference in the height of the *Tides* on the eastern and western coasts in those latitudes, the advantages resulting from this difference in a canal navigation.

9. Reflections on the alterations that would take place throughout the *West-Indies* and the country bordering on the *Gulf of Mexico*, if a wide canal or strait was formed as contemplated: would it not effect a diminution of the waters in the *Spanish Main* and *Mexican Gulf*, check the *Gulf stream*, lower the *Mississippi* bed, and occasion a disiccation of the shallow lakes, ponds and marshes in the low grounds? The advantages resulting therefrom to the city of New-Orleans and the surrounding country.

A concise view of the effects that a completion of the foregoing project would have on the political and commercial situation of the United States, as also that of other nations.

A brief elucidation of the above subjects and considerations will compose some future numbers, to be prepared and published when convenient.

In compliance with my proposition to the public, as published on the 8th ult. I will now proceed to examine the problem of the communication between the two seas, by taking a view of the several points that seem to offer the greatest facilities for canal or interior river communications; they are five in number, and comprised between the latitudes of 3 and 18 deg. N. We shall range them according to their geographical position, beginning at the most northern.

The Isthmus of Theuentepec to the south-east of Vera Cruz, which is 45 leagues in breadth from the Atlantic to the South Sea. The approximation of the sources of the river Huascualco, and Chimelepa, seems to favor the project of a canal or interior navigation; which would only require a length of six or seven leagues; a good road completed in 1800, on this portage, has served as a commercial intercourse between the two oceans. The elevation of the interior would require many locks for canal navigation, and the rivers are too small for the admission of large vessels. During the late war with Great Britain, the Indigo Guatimala and other articles came by the way of this Isthmus, to the port of Vera Cruz. This neck of land is situated about 350 miles south-east of the city of Mexico. The project of a water communication across the Isthmus, occupied the attention of some of the late Viceroys of this Intendency; but the strong opposition made by the principal inhabitants of the city, prevented the prosecution of the project on the ground that the completion of such a work would destroy the commerce of the city of Mexico, situated in the interior, and in which is concentrated an immense rich traffic from Europe and the East-Indies, through the ports of Vera Cruz and Accapulco, and which is carried on the backs of mules one hundred leagues, the distance across from one port to the other. There is not even to this day a road for a wheel carriage.

2d. Continuing southeast 200 leagues, we come to the great lake Nicaragua, 170 miles in length from northwest to southeast, and about half that in breadth; and has by the river San Juan, a grand outlet into the Spanish main; this river or strait is about 90 miles long, and is navigable for vessels of considerable burthen; a traffic has long since been carried on through this communication into the lake for a dye wood called [probably after the lake] Nigcorogo; the narrow isthmus of only twelve miles from the west bank of lake Nicaragua to the bottom of the Gulf of Papogayo is all that is to cut, to complete a water communication between the Atlantic and Pacific oceans.

In viewing this subject, two important considerations immediately present themselves. What is the depth of the Lake, and what the elevation of the ground on the Isthmus? As to the former I am not informed, but presume that from the great size of the Lake, and it being the receptacle of the waters flowing from several rivers, that the depth would be sufficient. This lake is also connected at its northwest corner with lake Leon, which communicates with the Pacific by the river Tosta. As it regards the height of the ground on the Isthmus, the Cordillera seems to subside, or at least is completely broken and that no elevated ridge exist in this quarter; as appears by Dampier who says "the coast of Nicoya is low and covered at full tide. To arrive from Ralixo to Leon we must go twenty miles across a flat country covered with mangel trees." This same navigator says that the ground between Nicoya and the lake is a little hilly, but for the most part level and like a savana.

Humbolt informs us that "The old maps" the Spanish maps "point out a communication by water as existing across the Isthmus from the lake to the great ocean. Other maps somewhat newer represent a river under the name of *Rio Perdido* which gives one of its branches to the Pacific ocean and the other to the lake Nicaragua. But this divided stream does not appear on the last maps published by the Spaniards and English." The same author tells us "there is in the archives of Madrid several French and English memoirs on the possibility of a junction of the lake Nicaragua with the Pacific ocean."

The foregoing authorities are sufficient to justify the belief that no continued elevation exists so as to prevent the cutting of a canal. The ground in some places is a little broken, but the only eminences seem to be isolated pyramidical summits, probably protruded by some ancient volcanic eruptions, and would be easily avoided by cutting round. It may be necessary to observe that the Isthmus between the lake and the Pacific ocean from the gulf of Nicoya to Leon is at least one hundred and fifty miles long. Surely in that distance on a narrow neck of land only from twelve to thirty miles wide and generally level that some place might be found that would be convenient to cut a canal. And as the water is much higher in the Spanish main that in the opposite ocean, a current would be formed that would soon wear a wide passage through. But I shall treat more fully on this part of the subject in a future number.

3d. The Isthmus of Darien, the narrow part of which is near Porto Bello and Panama; this place so long celebrated as the most proper point to open a water communication between the two seas, has not yet been fully explored; although it is three hundred years since it was first crossed by the Europeans.

This being the narrowest part of the American continent, has always been the spot to which public attention has been directed, more especially as the two great harbors of Porto Bello and Panama lying contiguous to that point on each side of the Isthmus has been the depots of an immense traffic carried across this neck of land, which at this place does not appear 60 miles wide.

The elevation of the ground at this point has never been correctly ascertained. The principal Cordillera seems to run towards the west side, from the summit of which it is said the two oceans are discernable, which would require a height of about nine hundred feet, making a deduction of one half for valleys that may separate the central hills, there will remain 450 feet of an elevation, which would require many locks, and could only be navigated by vessels that would be unable to keep the sea.

From every information that I have been able to collect relative to this celebrated spot, that has for so many ages interested and excited the public mind, I am fully of the opinion that the idea of constructing a canal like a pass or strait, so that large vessels could pass without locks from sea, to sea ought to be entirely abandoned.

4th. Continuing south along the coast of the Pacific 90 leagues to Cape Corrientes, in 5 deg. 12 min. N. we find the small bay and port of Cupica, from which place in an easternly direction, for 15 or 20 miles the ground is level and proper for a canal, which could be connected with the river Napi at a high navigable point. This river flows below the town of Zitra, into the great river Atrato, which enters the Atlantic at the Gulf of Darien. It appears that the chain of the Andes is entirely broken here; but what gradual elevation the interior may possess, has not yet been ascertained.

The distance from Corrientes to the bottom of the Gulf of Darien is 50 leagues, to which we may add one third for the windings of the valley. The whole distance will not then be two-thirds the length of the New York canal.

5th. Continuing south for one hundred miles from Cupica, we come to the mouth of the river Noanama, also called St. Juan, situated in 3 deg. 35. min. N. latitude, the sources of which interlock in the valley of Raspadura, with the waters of the Atrato, the beforementioned river that enters the Gulf of Darien.

The neighboring inhabitants observing the facility with which those streams might be connected, dug a small canal in the ravine of Raspadura, by which means in wet seasons loaded canoes pass from sea to sea. This internal water communication has existed for thirty years past, unknown to the people of the United States, until a few years since; for this and much other interesting information, relative to the Equinoctial regions of America, we are indebted to the celebrated Alexander de Humbolt. From the mouth of the Noanama on the Pacific, to the estuary of the Atrato, where it joins the Atlantic through the Gulf of Darien, on a straight line, is 315 miles, at this point, as well as from Cupica to the Gulf of Darien, the ridge of the Cordillera must subside, or be completely broken by transverse valleys, as is evident by the passage of small vessels from sea to sea, without the use of locks.

———•———

No. III

The pressure of the trade winds on that part of the Atlantic ocean, laying within the tropics, has evidently produced great alterations on that section of the globe called the West-Indies. The origin of this powerful current of wind is produced from several natural causes, the principal of which are the following: The diurnal motion of the globe revolving rapidly on its own axis from west to east, more than 1000 miles in an hour, under the rays of a vertical sun, the air within the tropics is kept continually heated, consequently rarified; it expands and raises to the upper part of the atmosphere, while the cold air from the temperate zones rushes in to maintain the equilibrium, but from the motion of the globe from west to east, those currents of wind have an apparent tendency from east to west; hence it is said that the winds follow the course of the sun. Although it is evident that the trade winds within the tropics have their apparent motion from the diurnal rotation of the globe, and not altogether from the action of the sun, as is generally alleged. It is true that the heat from that luminary rarifies the air as aforesaid; but as the sun is stationary, so would be the air as it regards any general current from east to west. But when we consider that that part of the Atlantic ocean near the equator, passes from west to east under the air that may be stationary, about 17 miles in a minute, will not the air appear to move rapidly, though it may remain stationary as it regards the space it is in? Thus the waters on the surface of the Atlantic that are near to the equator, are carried rapidly against this weight of air or wind that presses on the bosom of the ocean. The natural progress of the water from west to east is arrested, and, as it were, rolled back towards the west; this first impression or check is much assisted by the south-east trade winds that blow almost continually from the latitudes of 30 and 33 degrees south, in the Atlantic to the equator, as well as those that blow from the north-east, as high as the latitudes of 26 and 30 degrees; these north-east *trades* assist those from south-east in accummulating and heaping up the waters towards the equator, when both these currents of wind unite and blow steadily from the eastward, driving the whole mass of water on a surface of at least 500 miles in breadth, pressing on towards the eastern coast of the American continent, which it strikes with the greatest force near the *equator*; as the coast in that quarter turns towards the westward, the great body of this current takes its direction along the coast towards the windward islands, receiving on its way all the waters brought against this coast, by the trade winds from the east & north-east, as well as all the great rivers that flow from the American continent, amongst which is the *Amazone* or *Maragan*, the largest in the world, and the great Oronoco; these mighty rivers roll an immense body of water along with the general current, driven westwardly with the coast, between it and the windward islands. The first it meets is *Trinidad*, 90 miles long by 60 wide, the windward side of which is completely defended and secured by solid rock, the only reason that has prevented it from long since being swept away. Here the high land that once evidently joined this island to the continent at some ancient period, has been torn asunder by some tremendous convulsion of nature; the immense fragments of rocks that have been torn out and rolled away by this irresistible current, leaves no doubt of the reality of such an event. This current, with what passes north of Trinidad, is pent up in the Caribian sea and Spanish Main, and is strongly forced against the coast of *Costo Rico*, until the waters become accumulated at least 12 or 15 feet above the level of the opposite ocean. Following the windings round the bay of Honduras, to cape *Catoche*—still being pressed by the united currents of both wind and water from the south-east, this accumulation of water is driven through between that cape and cape *St. Antonio*, or the western part of the island of Cuba, into the Gulf of Mexico.— From the similarity of the ground and lowness of the land at cape *Catoche* and at St. Antonio, on both sides of the strait that divides them, not more than 60 miles wide, and the *soundings* that are found across the channel, induces a belief that the continual and incessant pressure of the waters broke down the barrier that connected the island of Cuba and the peninsula of *Yucatan* on the continent, and found a passage into the Gulf of Mexico, [*or rather into that once level tract of country that is now covered with that basin of water.*] Had not this event taken place, by which means the water found a passage by the point of Florida, a continued accumulation of that element, and being confined on the north by the island of *Cuba, yet* 700 miles long, and the large island of *St. Domingo* nearly adjoining, must inevitably found a passage over some level part of the American continent. Two points of that description might be found, to wit: at lake *Nicaragua* and near the gulf of *Darien*.

Pursuing this stupendous current into the gulf of Mexico, where it is joined by the waters of a number of large rivers flowing therein, especially that immense body of water that is continually poured from the mouths of that *Father of rivers, the Mississippi.* The water thus accumulated much higher than any part of the northern Atlantic, presses on towards that point, where the least resistance is found to oppose its progress, which is eastward, between the coast of Florida and the island of Cuba, but meeting in front with the extended banks and islands of Bahama, is turned north between these banks and the eastern coasts of Florida. Here this current first assumes the name of the *gulf stream*, and is reduced to the breadth of 40 miles wide, the *velocity* is at the rate of five miles an hour: this current continues to widen and runs almost parallel with the coast of the United States, deviating a little to the east of cape *Hatras*, and then running at the rate of two miles an hour, it strikes the banks of *Newfoundland*; here it meets with a resistance occasioned by that immense bank of sand, that no doubt has been worked out of the *Gulf of Mexico*, and carried to this place, where it originally met with a current from the north, an eddy was formed and the heavy *particles* contained in this once *muddy stream* was deposited there. This bank has increased to 400 miles in length from north to south, and 100 miles in breadth, and turns the current to the E.S.E.; this is occasioned by the bank aforesaid, together with the north-west winds that blow from the north-west parts of the American continent—and also, *a current* that comes down *Davis' Straits*, which originates from an annual *effusion* of the polar ices, and serves to keep up that necessary equilibrium in the ocean that otherwise would be lost by an excess of evaporation within the tropics. This current runs east of Newfoundland and the Fishing Banks and brings with it those *fields of ice* that are frequently seen near the Grand Banks in the early part of summer. This northern current joins the Gulf Stream to the south-east of the bank, in or near the latitude of forty-one, and is a great accession to its strength and velocity. Thus united, presses forward towards the *Azors*, or Western Islands. This current is felt by all the vessels that are bound from the northern parts of the United States to Madeira or the Canaries, that sail in the parallels of the Azors. This current continues on the north side and near to the island of Madeira, and passes on towards the coast of Africa, in a south-east direction, at the rate of one mile an hour, and strikes that coast with the greatest force in the latitude of 29 degrees north; it is then attracted southward and runs with great velocity along the coast, to supply the vacuum occasioned by the trade winds that first produced the origin of this *stupendous stream*, that has no parallel on the surface of the globe. Thus is the fountain constantly supplied by that immense mass of water that rolls round a circuit of 17 thousand miles in about twelve months.

It might be expected, that from the extraordinary pressure of the *trade wind* on the Atlantic within the tropics, in a distance of 3000 miles, and the prodigious effects resulting therefrom, that consequently similar causes would produce like effects in proportion to the extent of its operation, but it seems that the comparison will not always hold good; for the extent of the *Pacific* and *Indian* oceans together is about 13,000 miles in breadth from east to west, immediately within the tropics & is an extended & unbroken sheet or expanse of water, (except islands) from which we might reasonably expect the current would be stronger, and consequently the effects produced would be greater when it would meet an opposition; but the current in the Pacific and Indian oceans is not so strong as that in the Atlantic, for the following reason: The long and extended coast of southern *Asia*, lays parallel and contiguous to the tropic of Cancer in the northern hemisphere, where those variable winds called in that quarter the *monsoons*, that blow one half of the year from north to south, from the coast to the Indian ocean, and the other half the contrary direction, by which means the general current from east to west, occasioned by the *Trade winds* as aforesaid, is much broken, but notwithstanding it strikes the eastern coast of Africa, with considerable force, and drives the waters up northward into the *Straits of Babelmandel* and the *Red Sea*, so that in the latter they are much higher than in the *Medeterranean*, as was found when *Bonaparte* had possession of *Egypt*: It seems he intended to clear out the old *canal* that the ancient kings of that country had constructed; to form a communication between the *Nile* and *Red Sea*, he sent his engineers to examine the difference in the height of the two seas; they found that the Red Sea was 88 feet higher than the Mediterranean.

Thus in this case was the fact demonstrated, that the trade

winds have accumulated and raised the waters on the east side of the American continent, or that part comprised within the *Torrid Zone.* Is it not good reason to suppose, that the same cause will produce similar effects in every quarter of the globe where its operation can be equal?

------•------

NO. IV

The next consideration, that seems to present itself agreeably to our original arrangement, is the difference in the height, or rising of the tides, in the Spanish Main on the east side of the American continent, and the opposite ocean.

Humbolt tells us that the tide rises at Porto Bello, only thirteen inches, while on the opposite side, it rises from 13 to 16 feet. The enquiring mind will ask what are the causes that produce this difference, where the neck of land that divides the two seas, is not more than fifty miles wide?

The following short solution will serve to explain the question, and give the reason why; the water in the Spanish main is already raised, and accumulated nearly as much as the resistance of its gravitation will admit, which prevents the tide from the ocean from rolling up upon a summit, already as high as the tide could raise it.

This induces a belief that the acquired elevation of the water in the Spanish main, is equal to the height of the tides in the opposite ocean.

We will now proceed to examine into the advantages, that the elevation of the water, on the east side could, and would offer, in opening a wide canal or strait across the continent. Should a small canal be cut across, the bottom of which should be 6 or 8 feet below the surface of the water, on the east side, which, when let in, would soon wear out a wide passage. I know that much would depend on the different kinds of strata, of which the earth might be composed. But am of the opinion that nothing but a solid and extended rock, would be able to resist a current, driven and descending from such an immense and elevated mass of water, as the Spanish main, which is strongly forced against this coast by the trade winds as before described. Therefore, if a small canal was cut across some level part of the continent, the current would soon force a wide passage through.

Then all the waters, that are driven against this coast, from the east, and now forced up northward into the gulf of Mexico, would continue its natural course westward, down into the opposite ocean, which is at least from 12 to 16 feet lower. This current would then continue on through the wide Pacific.

Here an enquiry might arise how a vessel could return from west to east, against the force of the stream, that would flow into this new strait?

I answer, should the strait be worn to any considerable breadth, the water on the east side would become lower, and could not be expected to be higher than the rising of the tides on the west side, which would flow into the western extremity of the channel or strait: Thus would the vessel be carried back with the flowing of the tide from the west.

Is it not evident, that could there be a wide passage formed across any part of the continent, opposite to the Spanish main or Bay of Honduras, that the waters in those seas, would become much lower? Consequently a diminution of the water in the gulf of Mexico must follow.

This may be the proper place to remark on the alterations that would take place throughout the West Indies, and the country bordering on the gulf of Mexico.

The coast of the continent, and islands would be enlarged; the shallows that seem to connect some of the windward islands, would be laid bare, the numerous and extended islands, and Banks of

Bahamia, stretching in a direction from Turk's island, on the south east to the north west, for 700 miles, to the gulf stream, in the latitude of 26 degrees North, would be much increased. Many of the harbors in the West Indies, which are not at the mouths of rivers, would have their bars left much shallower: It would be difficult to approach the south side of Cuba, and the east side of the peninsula of Yucatan. The port of Vera Cruz, now the place of great resort for European vessels, especially the Plate fleet from Cadiz, would be much injured. It is even at this time, but a bad landing place, amongst shallows and rocks. But the harbors at the mouths of great rivers, that are already choked up with sand bars, occasioned by the conflict that always exist between the current of the rivers, and the rolling of the sea, at which point the heavy particles are deposited will in the event of the sea becoming lower, be forced away.

These remarks will be peculiarly applicable to the Mississippi. The prodigious whirlings of that great current of water, that passes cape Catoche by the peninsula of Yucatan, and rushes on towards the north west parts of the gulf of Mexico, which it strikes to the west of the Mississippi; one part of which turns west and south towards Vera Cruz, the other turns east, to the mouth of that river, and carries with it the sands that have been worked up in its passage; when this current comes to meet with the great body of water borne down by that river, the heavy particles are deposited there, is the reason why the bars at the mouth of the Mississippi are shallow. But in the event of the water being let through the continent, from the Spanish main, and the strong probability of its wearing through a wide passage, in proportion to the extent of that passage, a diminution in the height of the waters in the Spanish main, and Bay of Honduras, would take place. The current would no longer pass from those parts towards the gulf of Mexico; but would follow the contemplated channel to the south sea.

The gulf of Mexico would then be no longer fed from the

south, as at present. The gulf stream would soon draw off the excess of its waters. When done that stream would measurably subside; and the whole gulf of Mexico would in all probability, become eight or ten feet lower. Then no conflicting currents would prevent the Mississippi's stream from clearing its mouth of those sand bars, that now prevent large vessels from entering that river, and ascending as high as Natchez.

The same reason that prevents the tides from rising in the Spanish main, operates in the Mexican gulf; it is the extraordinary accumulation of its waters: but if these should subside, there is no doubt but the tide would rise in proportion to such diminution; and as the bed of the Mississippi would become lower towards the mouth, the tide would set up as far as that depression would take place.

But this flow of the tides up that river could not be expected at the time of high floods. If the channel of the Mississippi was deeper towards the mouth there would be less danger of its banks overflowing, and the great expense of throwing up, and keeping in repair, those dikes or banks on its verge, would be saved; there would no longer be any danger of the low country being inundated. The water would sink from the surface of the earth, in proportion as the gulf of Mexico and the Mississippi would become lower. A discication of the shallow lakes, ponds, and marshes would soon follow.

The inhabitants would no longer be compelled to breathe those humid and noxious vapors, that are exhaled by the heat of the sun from the surface of those stagnant waters that cover a great portion of this country. This terraqueous territory would become Terrafirma. The face of the country would assume a new appearance: N. Orleans, & its vicinity, from being the most sickly, might become one of the most healthy places; & the surrounding country would then be capable of supporting ten times the population. For, it is now more fit for the residence of amphibious animals, than the habitation of men.

As the Mississippi is the principal channel, through which the people of the western country can approach the sea with their exports, the commercial importance of New-Orleans is closely interwoven with our interest. *Query*— Will not the completion of our project render New-Orleans one of the first commercial cities in the world? As it is but a few days sail from the place of the contemplated canal to the mouth of the Mississippi, Orleans would be the nearest part in the United States that vessels from the north-west coast, and the East Indies could arrive at.

The difficulty that has hitherto existed in the passage from the Balize to Orleans is now overcome by the use of steam boats, towing vessels up. In proportion as Orleans becomes the place of deposite for our numerous exports, so in that proportion, will the commerce of Chili, Peru, the north-west coast and the East Indies be extended to it, as well as that of the West Indies, that lie so convenient.

The difficulty, that has heretofore existed in ascending the Mississippi, is now completely obvaited. The steam boats perform that voyage with surprising facility. So that places situated on the Ohio river seem to enjoy the advantages of a sea port.

———•———

V.

In viewing this subject deliberately, we are compelled to believe that the completion of the foregoing project, connected with the late improvement and important use of steam boats, would give to the people of the western country, important and lasting commercial advantages, such as the united energies of the world could not otherwise afford.

But it is not the western country alone that is to be benefitted, the Atlantic states would derive great commercial advantages. In distance, it would save 10,000 miles. Our trading vessels could then conveniently visit the north-west coast; and in a few months procure a sufficient quantity of fur, whale oil and seal skins, to purchase a cargo of east India productions. This kind of traffic would completely prevent and render useless the exportation of specie from the United States to the East Indies, that gulf, that has heretofore swallowed up the precious metals of every age and nation, and from which none returns.

It is true that some of our trading and fishing vessels have gone round cape Horn; and, on the northwest coast, have procured the above articles, which they have bartered in the East Indies for the production of those countries. But the long and tedious circum-navigation of South America is sufficient to discourage the most persevering adventurers.

The passing round Cape Horn, from east to west, is the most dangerous and difficult of any known to navigators, and can only be performed with probability of success in the months of December, January and February. It is then summer in the southern hemisphere.

I shall here entreat the reader to give his attention, while we take a short view of a passage from our ports to the north-west coast, by Cape Horn. Some navigators, who embark at our ports for the south sea, and who wish to land at Brazil or La Plata, cross the gulf stream, and make sufficient easting so that standing south they can clear the eastern side of the West Indian Islands; thence by Brazil and La Plata towards the Falkland Islands, between which, and the continent is the straits of La Mair through which they pass, but the great difficulty is to keep off the lee coast of Patigonia, and Terra del Fuego, on this dreary and inhospitable coast, vessels have been frequently wrecked.

But for half a century past the American navigators have preferred crossing the Atlantic in a south-east direction to strike the equator near the meridian of 20 deg. west of London; from thence a south-west direction beyond the tropic of Capricorn, where they meet the variable winds, that prevail in that quarter—still standing south-west so as to keep the Falkland islands to the west, and try to double Cape Horn in the latitude of 56 degrees, but here the wind from the west and south-west blows almost continually with a force, little less than a tornado; so that vessels bound west-ward, in this cold and inhospitable region have been compelled to beat against the storm for sixty days and nights without gaining an inch, and have some times been driven back and unable to perform the voyage that season. But to return to the point of difficulty and danger, where our brave seamen are literally struggling for their lives, with a perseverance and fortitude unknown to any other class of men. After weeks of incessant exertion in this unequal and tremendous contest with the infuriated elements, until hope (the seaman's sheet anchor) is about to fail, but from despair they derive courage, their exertions are renewed, the struggle is continued perhaps many days longer; while if possible, the danger seems to increase. The surging billows, over which they ride, seem ready every moment to swallow them.

Behold them tost [*sic*] upon the summit of a mountain wave, from which elevation, they are in a moment to be precipitated into the deep the watery and yawning gulf beneath, from which perhaps they never rise. But if providence in his mercy should rescue them from this untimely grave, and they be enabled to ride out the storm, double the Cape, and set their faces to the north, they have then 4,500 miles to sail; in most which distance, they have to struggle with currents on the coast of Chili and Peru, before they can return to the latitude of the contemplated canal.

The danger, the difficulty, and length of times necessary to perform this voyage, renders the number of adventurers comparatively few.

Could a passage across the continent be completed; from the

nature of our government, the commercial enterprize of our people, the number and excellency of our seamen, the United States would derive more advantage than any of the nations of continental Europe. For example, the New-England states with a population of 2,500,000 can produce more able seamen than France with a population of 30,000,000.

We will next proceed to examine what are the political advantages that the United States would derive from the completion of our object.

The convenience and facility with which the north-west coast could be approached, would enable the general government to plant and protect commercial colonies at, or near the mouth of Columbia river, or upon the whole extent of that coast, claimed by the United States.

Those immense and fertile regions of the west are; at present, much exposed to the rapacious encroachments of the Russian government, that mammouth of Empires, that seems ready to swallow up the balance of continental Europe, has for more than thirty years past, made constant and progressive approaches from the straits of Behring, in the latitude of 66 degrees north, round the peninsula of Alaska, continuing on the coast south-ward to Nootka sound, in a latitude of 48 degrees, a distance of 1,200 miles. But it seems their claim to this extent of the north-west coast is not yet sufficient.

We believe that a negociation has long since been pending between the courts of St. Petersburg and Madrid, for a considerable extent of territory on the west coast of America, from or near the entrance of the gulf of California, to the northern Spanish posts in the latitude of 37 degrees; a distance of at least 800 miles.

The late appearance of a Russian fleet on the American coast corroborates the fact.

I would ask the statesmen if these movements do not indicate an encroachment on our territory? If so, it must be left to the prudence and energy of our government to adopt such measures as may be necessary for the protection of that invaluable territory. But it is alleged that if a passage across the continent was formed, that the commercial nations of Europe would participate in the advantages resulting therefrom.

To this I am ready to accede. And they, perhaps, would assist in effecting it, more especially as a general peace seems to prevail throughout Europe at present.

But it is not only the commercial and christian world that would be benefitted. The heathen and illiterate nations, that inhabit the numerous and extensive islands (that lie scattered throughout the wide Pacific) of Australasia and Polynasia, together with four-fifths of the population of Asia, that are idolators, and which, together, compose more than one half of the inhabitants of this globe, would become enlightened. Literature, arts, sciences, and religion would be introduced amongst them.

A late celebrated author, who makes some remarks on this subject, says, "should a canal or communication be opened between the two oceans, the productions of Nootka sound and China would be brought 2,000 leagues nearer to Europe and the United States. Then only can any great changes be effected in the political state of eastern Asia; for this neck of land, the barrier against the waves of the Atlantic ocean, has been for many ages the bulwark of the independence of China and Japan." The same author speaks of it as "an undertaking calculated to immortalize a government engaged in the true interest of humanity." It could not be expected that our government would wish to make conquests in those distant regions. Commercial advantages, and the pleasing satisfaction of introducing civilization amongst numerous savage nations, that compose a considerable portion of the inhabitants of those countries, would, I trust, be a sufficient inducement for the undertaking.

Our age and nation would shudder at the idea of renewing and visiting upon those feeble people the horrid calamities their ancestors suffered by the first Europeans that visited the East-Indies, by the way of the cape of Good Hope. The most wanton destruction of the natives was the sport of the conquerors; the course of whose march was marked with fire and sword; while the ensanguined plains of India were moistened with the blood of her defenceless inhabitants. Vast pillars of smoke ascended from the venerable piles of her ancient pagodas. Many of those who survived the destruction of their temples and the conflagration of their country, were condemned by the *unholy* fathers of the horrid inquisition, to dance the *auto defee*, because they could not believe in a system of religion they had never before heard of!

But we trust those days of superstition and religious intolerance will no longer disgrace the conduct of civilized nations.

Should the completion of our project be effected, it would become the grand channel through which a large portion of the commerce of every civilized nation would pass: but none would be so much interested as the United States. Our citizens would be enabled to participate deeply in the advantages resulting from a traffic throughout the numerous and extended coasts bordering on the wide Pacific. It would be the *Tarshish* and the *Ophir* of modern times. Almost every sail would be bent in that direction.

It would be the grand and unprecedented march of commerce, of arts, of science, and of religion, that could, with facility, be waited across the Pacific. Our numerous trading vessels at Canton or elsewhere, protected by a sufficient force from the line, would command a respect and obedience, that princes, in those countries, would not dare to withhold; the odious, the useless and the tyrannical restrictions that are now imposed on the commerce of christian nations, would soon be removed.

The lowering clouds of ignorance and superstition, that have, from time immemorial, enveloped these oriental regions, would be dissolved. The minds of the natives would become enlightened. At that moment they would view with surprise and contempt the ridiculous fetters that had heretofore enchained both body and mind.

The iron grasp of Asiatic despotism would be unclinched.—The priest's magic wand would drop from his hand. The radiance of light would shine through every crevice of deception.

Their gods and their altars would literally tumble together in the dust; and on their ruins would be planted the tree and the standard of civil and religious liberty, entwined with the ensign of meek christianity.

The soul, thus enlightened and emancipated, would then go forth in the plenitude of his unbounded freedom and rejoice that he had become an intelligent being.

I shall close my remarks on this subject at present, by enquiring into the prospects our government could have in procuring the consent of the Court of Madrid, to enter upon her territory, examine the ground, and construct a canal. As the Spanish government has repeatedly rejected similar applications made by her own subjects, there is little doubt but she would be unwilling (if she could help it) to grant that privilege to others. But when we consider that the Spanish government holds the territory in question by a very precarious tenure, they might be willing to barter it for a small consideration, while any thing could be got for it. Should this province fall into the possession of the Patriots, we might expect to meet in them reciprocal and corresponding views.

But should all these considerations fail—we know what can be done.

Would it not be an easy matter for some of our public armed vessels, when in the South sea, to land on the narrow part of the Isthmus, between the Gulf of Papogoyma and Lake Nicaragua, which Isthmus or neck of land is only twelve miles across. The ground could be examined, and it, as alleged, that no elevated ridge exists, could not a few hundred men in two or three months let the water through by a small canal? If the bottom of Lake Nicaragua lies a few feet lower than the surface of the Gulf of Papagoyna, there is no doubt but nature would soon perform what the hand of man could not.

The foregoing project, with my views on the subject, is respectfully submitted to the American republic.

JOSEPH FOOS.

At some future period, not far distant, I expect to resume the above subject, and in the mean time solicit those of my fellow-citizens, who feel interested, to transmit to me, at Columbus, Ohio, such information as will throw light on this interesting subject.

J.F.

CANAL TO THE PACIFIC OCEAN.

MR. EDITOR,

Seeing in your paper of yesterday, some observations on the project of a passage to the Pacific ocean, by means of a canal across the Isthmus of Darien, brings to the mind a proposition made to the King of Spain, by Sr. Don Salvador St. Martin, the bishop of Chiapa, who resides at Ciudad Real, which is the capital of the province. West of this city, the river Goazacoacos takes its rise, and running east, empties into the Gulf of Mexico, about thirty leagues west of Vera Cruz; it has ten and twelve feet water on its bar, and is navigable for craft of four or five feet up to the above city. East of said city, the river (or a stream which empties into the river) Chimilapa takes its rise, and taking a westerly direction, empties into the Pacific ocean, at the port of Tahuantipa, which is a good harbor for large ships, and the river is navigable for craft of four or five feet, up to the above city of Ciudad Real: the two rivers passing each other at or near the city in a parallel line, at the distance of seven miles from each other; neither of them have falls to impede their navigation, and the ground through which a canal would have to be cut to connect the two rivers, is neither mountainous nor rocky.

I have seen a minute description of the rivers, and the countries through which they pass, contained in a petition to the king of Spain, begging permission to cut the above canal; it was presented in 1816, and contained the most convincing evidence of the facility with which the two oceans might be connected. The petition met with an utter denial from the king of Spain. Had it been granted, the intention of those concerned was, to have steam-boats employed between the two oceans, which would have made a voyage in much less time than it occupies between this and Louisville; and as the mouth of Goazacealcos is not more than eight or ten days sail from the mouth of the Mississippi, may we not with propriety hope that Spanish America will soon shake off the European yoke, when she will be at liberty to make improvements for her own benefit without consulting those who had no rule of government but their own jealous fears? Should such a communication be opened between the two oceans, what calculations can realize the future grandeur of New-Orleans?

[New Orleans' paper.]

HIGHWAY OF ALL NATIONS
PASSAGE ACROSS THE ISTHMUS OF DARIEN

Baron Humboldt offers nine points which have each been suggested as suitable points from which a canal across the Atlantic to the Pacific ocean could be made. Geo. Abercrombie made some minute surveys on the same subject some years ago, and laid his calculation and estimate of the labor before the British ministry; but nothing has been done. It is supposed that in case an independent government could be established in Mexico, the project will be revived. The waters of the Gulf are said to be considerably higher than those in the Pacific ocean, owing to the trade winds, which, blowing from the east, heap them up and force them to escape through the straits of Florida, thereby occasioning what we call the Gulf stream.

By cutting a passage across the Isthmus of Darien, or rather through a flat country, between the head of Nicaragua in 12 deg. north latitude, says Dampier, and the coast of Nicoga, where there are no mountains, would make only twenty miles across a savanna covered with trees; then the waters would rush thro' the opening, and by degrees wear a fine and wide channel, till the two oceans would become nearly on the same level.

The navigation to the East-Indies would be shortened near ten thousand miles. The waters would recede from the coasts all around the gulf, and increase the territories of the bordering countries.

The West-India Islands would grow every day while the channel was wearing. Mariners would no longer go by the Gulf stream from Florida to Newfoundland. Let all the nations in the world who are interested in accomplishing this object, make a joint attempt, and the work would soon be completed. We hope statesmen will reflect seriously on this point, which is bro't to our remembrance by a resolution of the Ohio Legislature.

[Nashville Clarion]

APPENDIX

BIOGRAPHICAL MATERIAL

Joseph Foos was born in Chester County, Pennsylvania, about 1767. His father was a native of Germany and his mother of Wales. He removed with his parents to Tennessee, and afterwards to Harrison County, Kentucky, where, in 1797, he married Lydia Nelson, and the following year moved to Franklin (now Columbus), Ohio, where he owned the ferry over the Sciota, then a franchise of great value; he also kept a large tavern. He was a man of fine natural ability, and though he spoke German and English with fluency and elegance, his early education was deficient. His memory was remarkable, and his perception very quick; and after taking private instruction of an Irish schoolmaster, who came to his house in indigent circumstances, he came to be regarded as a man of more than usual acquirements, and throughout the remainder of his life carried on a voluminous correspondence with Clay, Ewing, Corwin, Harrison and other contemporary characters of prominence. He was a member of the first Legislature, and before his death served twenty-five sessions in the Senate and House. He became an eloquent and moving speaker, and it was mainly through his persistent efforts that the capital of the State was removed to Columbus. In recognition of the services the authorities of the city afterwards gave him a square of ground, allowing him to choose it for himself. He served in the War of 1812, rising from the rank of Captain to Brigadier-General. During the years of this war and the Indian troubles that followed, Franklin was an important military post and his tavern the resort of the army officers. His opportunities for making money were very great; his ferry alone, during the movement of military forces and the tide of emigration sweeping in great caravans to the plains of Illinois, frequently netted him three hundred dollars a day. But his liberality was equal to his resources. His house was the head-centre for political agitators, and they were always needy. Even in entertaining such men as Clay there was more distinction than profit. His influence throughout the State at this time was undoubtedly very great, but it suffered a decline. He was defeated for Congress, and his property having depreciated by the changed circumstances of the country, he removed to Madison County and engaged in farming. About 1825 he was appointed General-in-Chief of the State Militia, and held the office until his death. He had taken a remarkable interest in the study of geography, and when the subject of canals was agitating the country, after the inception of De Witt Clinton's great scheme in New York, his attention was drawn to the feasibility of a ship canal across the Isthmus of Darien.

He opened correspondence with the Spanish authorities, who were civil enough to furnish him the required information in furtherance of his plan for a grand highway of nations; he furnished a pamphlet with a map. While this was remembered it was known as "Foos' folly," but several years after his death it reflected great credit on his name, when some controversy arose between England and the United States on the subject of the discovey of the route, and Tom Corwin arose in Congress and drew attention to the fact that the idea had originated years before with a citizen of the State of Ohio. General Foos' first wife died in 1810, having two sons and two daughters; and in 1812 he married Margaret Phifer, of Madison, with whom he had six children, five sons and one daughter. He died in 1832, and was buried at Columbus.

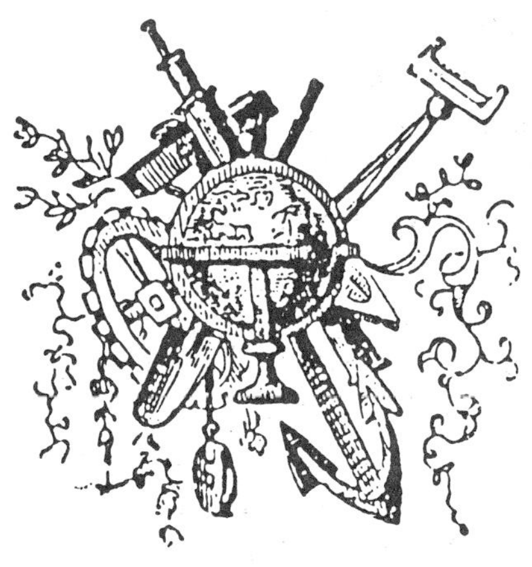

Index

The Joseph Foos pamphlet, THE HIGHWAY OF ALL NATIONS *was printed in Columbus, Ohio in 1820. After sleeping for 167 years it is here printed again in the workshop of Glen Adams, which is located in the quiet country village of Fairfield, southern Spokane County in Washington state and one township removed from the Idaho line. The booklet was set in type by Dale LaTendresse using a Model 7300 Editwriter computer photosetter. The typeface used is 14 on 16 Janson. Camera-darkroom work was by Sylvia Fenich using a 660C DS (Japanese) camera and a LogE automatic film developing machine. The sheets were printed by Dave Hooper using a 28-inch Heidelberg press, model KORS. Folding was by Garry Adams using a 22x28 Baum folding machine. Assembly was by the Ye Galleon crew. This was a fun project. We had no special difficulty with the work.*